MAINTAINING SMALL-FARM EQUIPMENT

How to Keep
Tractors and Implements
Running Well

Steve & Ann Larkin Hansen

Storey Publishing

The mission of Storey Publishing is to serve our customers by
publishing practical information that encourages
personal independence in harmony with the environment.

Edited by Deborah Burns and Sarah Guare
Series design by Alethea Morrison
Art direction by Jeff Stiefel
Text production by Theresa Wiscovitch
Indexed by Christine R. Lindemer, Boston Road Communications

Cover illustration by © Joe Wilson
Interior illustrations by Koren Shadmi

Storey books are available for special premium and promotional uses and for customized editions. For further information, please call 1-800-793-9396.

Storey Publishing
210 MASS MoCA Way
North Adams, MA 01247
www.storey.com

Printed in the United States by McNaughton & Gunn, Inc.
10 9 8 7 6 5 4 3 2 1

LIBRARY OF CONGRESS CATALOGING-IN-PUBLICATION DATA

Hansen, Ann Larkin, author.
 Maintaining small-farm equipment : how to keep tractors and implements running well / by Ann Larkin Hansen and Steve Hansen.
   pages cm
 ISBN 978-1-61212-527-5 (pbk. : alk. paper)
 ISBN 978-1-61212-528-2 (ebook) 1. Farm equipment—Maintenance and repair.
 2. Tractors—Maintenance and repair. 3. Agricultural implements—Maintenance and repair. I. Hansen, Steve, 1949– author. II. Title. III. Title: How to keep tractors and implements running well.
S675.5.H36 2015
631.3—dc23
                              2015019023

CONTENTS

ACKNOWLEDGMENTS

Both this book and our mechanical knowledge owe a great deal to the talented friends and mechanics we have known in our years of dealing with equipment. We'd especially like to thank Lawrence Reynolds III, Arlyn and Kevin Rediger, and Tim Stanton, along with the guys at Badger Sales, Swoboda Implement, Tractor Central, and Baribeau Implement. Last, Ann is still grateful to our old neighbor, Gary Bathke, who spent a summer's evening long ago installing a cut-off switch in the Farmall so that she didn't have to jump-start it every morning.

Any mistakes are, of course, our own.

Equipment amplifies our efforts so we can do more work with less time and muscle. A generation ago, farmers and home-steaders learned from their older relatives how to maintain and repair equipment, but few of us now have that resource. We hope this book helps to fill that gap.

Farm equipment falls into three general categories: human powered, small scale, and large scale. Maintenance and repair of human-powered tools requires little instruction; all that's needed is occasional cleaning, tightening, sharpening, and applying linseed or tung oil to the wooden parts. Large-scale equipment, from 100-horsepower-plus tractors to 36-row corn planters, is generally beyond the needs or financial means of the small-scale farmer.

Therefore, this book covers the engines and implements that fall between human and huge, the kind found on most small-scale farms, and which we own and have learned to maintain and repair through more than 20 years on our own farm.

A bad mechanic's favorite tool is a hammer;
and his second favorite tool is a bigger hammer.
A bad mechanic understands the true meaning of
"cobbled together," and his solutions are based on
colorful language instead of knowledge.
A bad mechanic believes "safety first"
simply takes too much time.

A good mechanic takes pride in his work.
He is systematic and stays organized;
maps out a solution before starting the trip;
anticipates pitfalls; understands how something
is supposed to work so that he can recognize any
problems; and asks for help when it's needed.

—Friend, farmer, and repairman Butch Reynolds

CHAPTER ONE

PARTS

The various equipment available to the small farmer is built from a fairly standard palette of part types. Understand the function of each type of part and see how the parts are assembled; you can then reason through how power is moved from engine to application. Learning how to keep equipment working follows logically from this starting point, so we'll begin by discussing the types of parts common on farm equipment — fasteners; wheels and shafts; bearings and bushings; and hoses and lines — and their functions.

Each type of part comes in a variety of permutations; for example, not all springs look like a coiled wire. If possible, read this chapter with a piece of equipment in front of you so you can get a real-life look at types of parts; illustrations and text are great starting points, but getting your hands on metal is necessary to really grasp how things fit together.

FASTENERS

Parts that hold things together are employed whenever you need to attach an implement, disassemble something to do general maintenance, or make a repair. The most common types of fasteners are bolts, rivets, clips, spring clips, keys, cotter pins, hose clamps, springs, shafts, U-joints, and hitch pins.

Bolts, Nuts, and Washers

Bolts are threaded metal shafts that are blunt on one end and have a head on the other. To hold them in place requires screwing a nut onto the shaft, or screwing the bolt into a threaded recess. **Nuts** are either standard or locknuts, which are designed to provide extra holding friction where there's constant vibration. **Washers** are used when a hole is bigger than the bolt head or nut, or to hold the bolt head or nut farther from the hole to shorten how much shaft sticks out the other side. Split or star-type washers have the same function as a locknut.

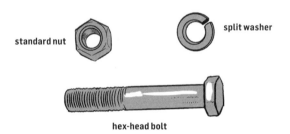

standard nut

split washer

hex-head bolt

Four Important Things to Know about Bolts

1. Bolts are sized by the length and thickness of the shaft and defined by thread count (coarse or fine) and head type: for example, a quarter-inch by two-and-a-half-inch (¼" x 2½") fine-thread hex-head bolt.

2. Bolt heads (and threads, nuts, and shaft sizes) are made to either metric or US measure, so you'll need either US or metric socket wrenches (see chapter 4 on tools) or, more likely, both.

3. Bolts come in varying degrees of hardness. Numbers 2, 5, and 8 bolts (from softest to hardest) are the grades most commonly used in machinery. Hardness is indicated by the number of lines on the head: no lines for grade 2, three lines for grade 5, and six lines for grade 8. Metric bolts are marked with the numbers 8.8, 10.9, and 12.9 for equivalent grades. When you need to replace a bolt, the new bolt ideally should be of the same hardness.

4. The compressive friction of the bolt head and the nut against the pieces they are fastening together is what gives the connection strength; in other words, loose bolts wear and break more easily than tight bolts.

Rivets

Machine rivets have smooth shafts, half-round heads, and a very low profile; they're used to hold parts moving through tight spaces. Removing and installing rivets is done with a riveting tool. You will run across rivets

rivets

mostly on older equipment. Since rivets can be fussy to replace when worn or broken, if there is room to substitute bolts it makes future repairs simpler.

Clips, Cotter Pins, and Keys

Clips of various sizes and shapes are used to hold and align parts where there won't be too much force and are designed for quick insertion and removal. The most common is the **hitch pin clip,** which is pushed through a hole in the bottom of a hitch pin (which connects an implement to the tractor) and keeps the pin from bouncing out.

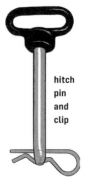

hitch pin and clip

Zerks

Also called a grease fitting, a **zerk** consists of a round head with a spring-loaded ball valve which looks like the tip on a ballpoint pen. It is attached to a hollow, threaded shank that screws into a port in the equipment wherever grease needs to be delivered to moving parts.

zerks

Cotter pins are a specialized type of clip, made of soft metal so they will break under unusual force and spare the more expensive parts they hold together. They're also used to lock nuts in place. They are intended to stay in place until they fail.

cotter pins

Keys are used to lock together a shaft and pulley or shaft and sprocket. Short canals or grooves are machined on the exterior of the shaft and the interior opening in the pulley or sprocket, and the key (either rectangular or half round — called a **half-moon** or **Woodruff key**) slips into the grooves and locks the two parts together. A **set screw** — a short, stubby screw that screws through a threaded hole in the exterior part to bear on the key or directly on the shaft — may be used as insurance to prevent the key or shaft from slipping out.

key

shaft

keys

Spring Clips

A **spring clip** is a slim metal bracelet with a cut in it that is slipped into a groove around a shaft to hold a rotating wheel in place on the shaft. On either cut end is a loop for the points of a **spring clip tool** (a specialized type of pliers), used to install and remove spring clips.

....

Spare Parts to Have on Hand

Available at farm equipment stores, and many hardware and auto supply stores:

- Selection of bolts, nuts, and washers (purchased or scavenged)
- Selection of rivets if any of your equipment has rivets
- Selection of cotter pins
- Selection of zerks
- Selection of chain links
- Extra hitch pins with clips
- Selection of hose clamps
- Selection of bushings and spacers (purchased or scavenged)

Available from equipment dealers or from the manufacturer:

- Spare belts (specific to each machine)
- Air and oil filters (specific to each particular engine)
- Spark plugs (specific to each particular engine)
- Parts that frequently need replacing on your equipment, such as rake teeth, cutter bar sections, cultivator sweeps, etc.

....

Springs

Springs hold parts together in a number of ways, depending on what's needed: flexible contact, reducing shock or impact, allowing connected parts to move somewhat independently of each other, or as part of a mechanism to engage and disengage two other parts. When a spring breaks or is distorted, it should be replaced with a close if not exact replica, so that its function is maintained.

WHEELS AND SHAFTS

WHEELS COME IN ALL SHAPES and sizes on farm equipment, from tires and sprockets to pulleys and gears. Wheels fit on **shafts**, which are long metal rods. The shaft between tires is called an **axle**; a short, tapered shaft is called a **spindle**. A spindle is a type of half axle which can be articulated to allow the tire to steer independently of the axle. A shaft that has a pulley or sprocket on it may also be called a spindle.

Tires

Farm equipment tires are built and treaded for traction, not speed. The rear tires on tractors are usually filled with a mix of calcium chloride and water to give them added weight for better traction.

Universal Joints

A **universal joint** (more commonly referred to as a U-joint) creates a flexible joint between two shafts. The ends of the shafts are U-shaped and connected by a metal cross with bearings at each corner where it attaches to each of the legs of the U-shaped ends. The bearings allow one shaft to turn freely side to side and the other to move freely up and down, allowing changes in the angle, all while the shafts are spinning in unison. U-joints should be greased diligently.

U-JOINTS ALLOW A ROTATING SHAFT to bend as it rotates; they are common on farm equipment. Keep them well greased and stay clear of them when they're turning: they can snag loose clothing or laces and twist, tearing fabric and flesh.

Pulleys and Sprockets

A **pulley** is a grooved wheel made to hold a belt, while a **sprocket** is a wheel with cogs made to hold a chain. The relative sizes of pulleys and sprockets determine the gear ratio, serving to increase or decrease power versus speed as required for function.

Gears

Gears are cogged wheels fitted tightly together in precise combinations to change the ratio of force to speed between engine and implement, and to change the direction of power, such as when it needs to go around a corner. For the most part, gears on farm equipment, though essential, are unseen inside sealed gearboxes and rarely need attention.

PTO Shaft

Where the power source is separate from the implement, a short, splined **power takeoff (PTO) shaft** — on a tractor, for example — inserts into a splined receiving shaft extending from the implement to deliver power to the working parts. PTOs have either six splines and turn at 540 rpm (revolutions per minute) or 21 splines and turn at 1000 rpm. The PTO shafts of implements have to match those on the tractor.

The short, protruding PTO shaft is probably the most dangerous part of any piece of equipment you'll own; brushing it with a loose shoelace when it's running can get your foot ripped off. Make sure there's a guard over it, and never get close to a running PTO (or a rotating U-joint).

Clutches

A **clutch** interrupts a driveshaft so a power source can be engaged and disengaged, which allows the engine to keep running while wheels and the PTO are not. Tractors with "live" PTOs (that is, all but the oldest tractors) have separate clutches for wheels and PTO. Clutch repair is generally a job for professionals.

A second type of clutch found on some bigger implements is the **slip clutch** (also called an overriding clutch), which is engineered to disengage under exceptional force, to prevent other parts from breaking. Slip clutch work is generally within reach of the average farm shop.

Belts and Chains

Belts and chains are common on all types of equipment, and they loosen, wear, and break fairly often. Both function by being wrapped around two wheels (pulleys or sprockets), one of which is turned by the engine or motion of the wheels, moving the belt or chain, which then turns the pulley or sprocket at the other end, which is attached to and turns a working part of the implement.

Rubber belts may have a V-shape in cross section and then are called **V-belts**, but there are other types, and it's important to replace a bad belt with another of precisely the same length and shape. Very old equipment may have leather belts; apply spray-on leather-belt conditioner to make these last. Chains must be sized to fit the sprockets; they're used where no slippage is wanted (belts allow for a little slippage).

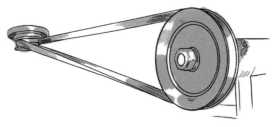

BELTS CAN CHANGE the direction and speed-to-force ratio of power. They rely on friction to turn pulleys, so they must be tight and free of grease and oil.

BEARINGS AND BUSHINGS

WHEN METAL MOVES against metal, it binds and grinds unless a fluid or mechanism or combination of the two is put in place to minimize the friction. Grease and oil are the fluids (see chapter 3); bearings and bushings are the mechanisms used to minimize friction.

Bushings are thick, relatively short metal cylinders made of softer, more slippery metal, used to provide space between moving parts. **Bearings** are more complex, precisely machined, and expensive.

Bearings allow wheels to turn independently of an axle, spindle, or shaft. On farm equipment, most bearings are made of little steel balls or cylinders that are packed in a bed of grease in a groove inside a metal doughnut-shaped "raceway" and capped with another raceway. The raceways move separately, the bearings rolling in between to prevent friction between top

A TYPICAL BALL-BEARING ASSEMBLY
consists of five basic, and tightly packed, components.

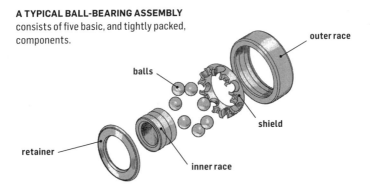

balls

outer race

shield

retainer

inner race

and bottom. Bearings are either sealed and need no regular attention, or unsealed and greased by disassembling and hand-packing, or through an external zerk (see page 6).

HOSES, LINES, AND PARTS SPECIFIC TO FARM IMPLEMENTS

Hoses and lines (small versions of hoses) are used to circulate fluids, including fuel, coolant, and hydraulic fluid. Coolant hoses and fuel lines are attached with clamps or fittings, while higher pressure hydraulic hoses utilize couplings, either screw-together or quick couplings. **Hose clamps** are perforated strips of flexible metal that cinch around a hose end where the hose fits over a connection. A small bolt holds the clamp tight, keeping the hose in place.

Parts that contact dirt, seeds, weeds, crops, and manure are specific to the job and, because of constant hard use, need to be replaced fairly often. Implements are designed to make this routine chore simple, and replacements are commonly available at farm stores and equipment dealers. Though the variety of parts in this category is huge, some of the most common are discs, sweeps, coulters, cutter bar sections, spring teeth, and rake teeth.

CHAPTER TWO

SYSTEMS

Systems are assemblies of parts that create, move, and apply power; move fluids; and cool and lubricate. An engine has five basic systems: fuel, electrical, lubrication, cooling, and air intake/exhaust. The system that moves power from engines to implements is a drivetrain. Small machines may make this connection by belt or chain. On large equipment, such as tractors and skidsteers, there is often a gearbox and/or a hydraulic system for moving power from engine to implement.

When dealing with equipment parts and systems, think in terms of "if this, then that." For example, if there's a zerk, then there's a bearing, bushing, or shaft. If there are hydraulic hoses, then there's a cylinder, pump, and couplings. If there's a wheel — whether it's a tire, sprocket, or pulley — then there's a shaft and bearings.

FUEL SYSTEM

IN A GAS-POWERED ENGINE, fuel is moved by gravity or a fuel pump from the tank through a line with either a filter or screen and into the carburetor, where it is mixed with air. Inside the engine block, the air-fuel mix is sucked into the cylinders, where it is compressed by the pistons and ignited by a spark from the spark plugs. The fuel explodes, driving the pistons back. Rods attached to the bottoms of the pistons turn the crankshaft, converting the energy of the chemical explosion into mechanical power. Note that air, fuel, and spark must all be present for combustion to take place in a gas engine.

A diesel engine has no carburetor or spark plugs; the fuel is injected directly into the cylinders, where it is mixed with air. Extreme compression of the mix causes ignition.

This very simplified explanation of the workings of an internal combustion engine completes the description of the fuel system and illustrates where the fuel system intersects with the electrical system, the air intake/exhaust system, and the drivetrain.

Two-Cycle vs. Four-Cycle Engines

Larger engines, whether gas or diesel powered, are **four-cycle**, so-called after the two-up plus two-down strokes of the pistons which complete the power and recharge cycles. The first down-and-up crankshaft rotation draws in and then compresses the fuel air mix. A spark delivered to the cylinder at peak compression ignites the mix. The resulting explosion powers the second down-and-up rotation, driving the piston down to turn the crankshaft, which creates momentum to move the piston upward to expel the exhaust gases and set up the next cycle. Small engines are gas powered and either four- or two-cycle.

Two-cycle engines tend to be smaller and lighter, but are less efficient. This type of engine utilizes just one up and one down stroke of the piston for intake, ignition, and exhaust.

For everyday purposes, the difference between the types is that four-cycle engines have a separate system for lubrication, while two-cycles add oil to the gas, which lubricates the entire engine. Since two-cycle engines don't require separate engine lubrication systems, you don't ever have to change the oil or an oil filter.

Very few two-cycle engines mix the gas and oil internally; most (chain saws are one example) require premixing the gas and oil before adding it to the fuel tank. Use the correct type of oil, and the correct proportions of gas and oil, as specified in the owner's manual. (Don't forget that a chainsaw has a

Valves

Valves are metering devices for liquids and gases, which operate inside a line or chamber that contains the fluid, forcing it to pass through the valve. They're found in several places in different systems. Hydraulic controls are valves, as are the choke and carburetor float valve in the fuel system, combustion valves at the cylinders, and the thermostat in a coolant system.

SCHEMATIC OF FUEL, AIR, AND ELECTRICAL SYSTEMS

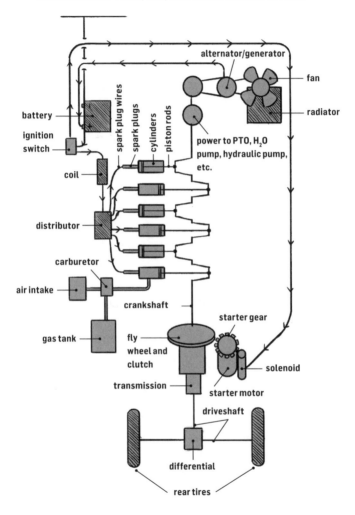

separate lubrication system for the cutter bar and chain, which uses a different type of oil than the engine.)

ELECTRICAL SYSTEM

THE ELECTRICAL SYSTEM STARTS the engine and powers some other components, such as lights (if any) and the instrument panel. When you turn the ignition key or switch to "on," it completes a circuit that puts electricity from the battery through the spark plugs. When you turn the switch a little farther to "start," it completes a second circuit from the battery to the **solenoid** (an electromagnetic piston), which engages the starter motor. This motor starts, and the solenoid jams the starter gear forward to engage the cogs on the flywheel. The flywheel's turning starts the pistons moving, creating suction in the cylinders, which draws the fuel-air mix in through the carburetor, where it meets sparks from the spark plugs. This ignites the fuel, the explosion forces the pistons down, the crankshaft begins to turn, and the engine starts running on its own. You then move the switch back to the "on" position, which disengages the starter motor.

On engines with no battery, pulling the starter cord spins the crankshaft and its built-in generator, the **magneto** (basically a fixed magnet and a wire coil), to create an electric current to the spark plug.

Once the engine is running, the battery is no longer needed as the power source for generating sparks through the spark plugs. That job is turned over to one of three devices — which one depends on the age and size of the engine: a magneto, a generator, or an alternator. All are a type of generator that provides energy (spark) to the spark plugs and simultaneously recharges any battery. These devices are powered by a driveshaft or a belt coming off the driveshaft, called the generator or alternator belt.

ENGINE LUBRICATION AND COOLING

AT THE BOTTOM OF THE ENGINE BLOCK on a four-cycle engine is the oil pan, which acts as an oil reservoir. Exposed engine parts are continuously bathed in oil either by the crankshaft's turning through the pool of oil and splashing it up, or by an oil pump that forces oil through a filter and then onto the engine parts. When oil gets low, the engine has to work harder against wearing friction, and when oil gets dirty it can gum up or scratch engine parts. *Checking the oil level daily and changing it regularly are the two most important things you can do to extend engine life.*

Since fuel ignition creates intense heat, engines must be cooled so the heat doesn't warp or melt parts. Small engines are usually air cooled; the cylinders have fins on them to increase heat loss. Bigger engines use liquid coolant — a mix of half antifreeze and half distilled water (see chapter 3 for more on this topic).

Coolant in the radiator is moved by the water pump through a series of cavities on the inside of the engine block, where it absorbs heat from the engine. It then circulates through a hose past a thermostat back to the **radiator**, essentially a set of small, finned tubes engineered to present maximum surface area for quick cooling.

On a large engine, the cooling airflow across the radiator is increased with a fan run by a fan belt. The thermostat is designed to open the line once the engine reaches optimal operating temperature. Because any air in the system expands as it heats, the cap on the radiator has a pressure valve to vent the air to release the excess pressure.

AIR INTAKE AND EXHAUST

FUEL WON'T IGNITE WITHOUT being mixed with air, and efficiency of combustion depends on having the right ratio of fuel to air, which are mixed inside the **carburetor** (a metering and atomizing device for mixing fuel and air). In a diesel engine, the mix is metered by fuel injectors. The **air intake system** delivers clean air to the carburetor by means of an air filter over the air intake or, as on many old tractors, a precleaner and oil bath in front of the intake. Air filters can be foam, paper, metal, or a combination.

From the carburetor, the intake manifold distributes the air/fuel mix to the cylinders. (A manifold is something that collects from one and distributes to many, or vice versa; in an engine, it is a pipe or chamber that branches into several openings.) The exhaust is pushed into an exhaust manifold, then piped through the muffler, which serves to muffle the noise of the engine, arrest sparks, and prevent cold air contact that might warp the exhaust valves.

POWER TRANSMISSION

Rods connect the pistons to a crankshaft inside the engine block so that when the pistons move it turns the crankshaft. To move this power to the working parts of a machine, the end of the crankshaft is bolted to the flywheel, which is mated to one side of the transmission clutch. The other side of the clutch is bolted to the **transmission**, a gearbox that governs the driveshaft, so that when the clutch is engaged and the engine is in gear it turns the driveshaft.

At its far end, the driveshaft powers the differential gear in the middle of the rear axle, which turns the axle and hence the rear wheels. The crankshaft also separately powers the PTO clutch and shaft; the oil, water, and hydraulic system pumps; and the alternator, generator, or magneto that provides electrical spark.

Brakes

On any riding-type motorized equipment there will be brakes, which are applied after the clutch is disengaged, to stop further forward motion. Brakes require little attention unless they're performing poorly. Old tractors are notorious for having bad brakes, but this sometimes may be poor adjustment rather than brake failure. Older brake systems are mechanical, while newer equipment may have hydraulically powered brakes instead of rods and cables.

Ground-Driven Power

Some small or very old implements are ground driven instead of being powered by the engine. The act of pushing or pulling the implement over the ground supplies all the power needed. The simplest example of this is the **drag**, where the friction of the implement — whether it's made of a frame with teeth, an old bedspring, or a bundle of brush — against the dirt accomplishes the purpose of smoothing the seedbed. All that's needed is for a person, an animal, or an engine to provide forward motion. Other ground-driven implements capture the motion of the implement's wheels with gears, belts, or chains to transmit to other working parts. A walk-behind manual planter in the garden is one example of this; another is the ground-driven side-delivery hay rake, where a belt connects the wheels to the rake reel to turn the reel.

The clutch, gearbox, driveshaft, and differential gear, which together are called the **drivetrain**, tend to be well protected by heavy metal, making them hard to get at. Problems with these components can be complicated to fix but fortunately are infrequent.

In addition to the transmission and differential gearboxes associated with engines, many implements have gearboxes as well, which change the speed-to-force ratio of the power delivered by the PTO or other source to what's appropriate for the implement, and can change the direction of power as needed.

Belts and chains are the final step in many implement power transmission systems and may be the only part of the system present in simple small engines. Belts and chains wrap around a drive pulley or sprocket at one end and a driven one at the other end. Pulleys or sprockets will be either integral to the shaft or axle or held with a key. Any tensioning sprockets and pulleys (for holding a chain or belt tight) are freewheeling and have bearings of some sort.

HYDRAULIC SYSTEM

HYDRAULIC SYSTEMS utilize the noncompressible property of fluids (hydraulic fluid in this case). The fluid is pushed from a reservoir by a pump through a system of high-pressure hoses, connected by metal hydraulic couplings, through valves that move the shaft on a hydraulic cylinder back and forth, which in

Transporting and Positioning Implements
Implements separate from engines commonly have both working and traveling positions. Newer implements usually allow you to adjust the position with a hydraulic connection from the comfort of your tractor seat; older implements may require you to get down and turn a handle or pull some clips and push things into position, then reinsert the clips to hold the new position.

turn moves an implement up, down, in, or out. Levers controlled by the operator open and close the valves. Hydraulic power is impressive, but on most farm implements it doesn't make things turn; it positions and holds implements and attachments.

Some larger equipment, such as skidsteers and some riding lawnmowers, utilizes hydraulic drives, which are powered by the combustion engine and replace the transmission, driveshaft, and differential gear.

Hydraulic couplings are either screw-together or quick couplings when they are frequently attached and detached. The brand of coupling on the implement hoses must match those on the tractor or they won't connect. Always wipe coupling threads and surfaces with a rag, glove, or hand each time you connect them; grit in the couplings leads to leaking fluid and bad connections.

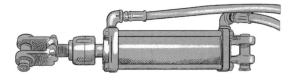

A HYDRAULIC CYLINDER has a moving shaft that can lift, lower, and hold equipment.

CHAPTER THREE

FLUIDS

Fluids are critical to keeping parts moving and systems functioning, and are the primary focus of a maintenance program. Engines require fluids in the form of fuel and air to operate. Both engines and implements require fluid lubricants — grease, lubricating oil, engine oil, and gearbox oil (transmission fluid) — to prevent unnecessary wear and breakdowns. Large engines additionally use coolant and hydraulic fluid. There is also a variety of fluid aids and additives for machinery and engines, used to protect or clean different systems.

GREASE

Most moving mechanical (non-engine) parts rely on grease for lubrication: a combination of oil, soap, and additives formulated to work through a wide range of temperatures and loads. There are hundreds of kinds of grease available, but the only kind needed on most farms is any brand of multipurpose, or MP, grease, either lithium or petroleum based. Still, check your manual to see if a different type of grease is recommended for a particular piece of equipment or a particular mechanism. Tubes of grease are available at any farm store, and grease is applied with a grease gun through zerks, or wiped onto parts with a finger or paper towel.

A LOADED GREASE GUN with a flexible nozzle is the most important tool on a daily basis for keeping equipment running.

LUBRICATING OIL

LUBRICATING OIL IS USED on parts, such as chains, that need less bulk and more fluidity in their lubrication than grease offers. It also acts as a rust preventive; so that keeping a light coating of oil on nuts and bolts helps ensure that when you need to unbolt things they will come apart easily. *Do not use oil on belts,* which depend on friction to operate. Use a light oil, such as WD-40, on chains and more delicate mechanisms. On heavy-duty parts exposed to the elements, heavier oil is better; you can even recycle old engine oil for this purpose.

If you're often running equipment in very dry and dusty conditions, or on very sandy soil, use a dry lubricant such as graphite or dissolved paraffin wax (White Lightning is one brand) instead of oil on chains, to keep grit from sticking to and wearing the chain. For nonmoving parts that must be absolutely dry to function properly, such as the seed bins on planters and drills, use a dry lubricant or a light coat of spray-on metal paint as a rust preventive. When bolts are frozen in place, a penetrating oil such as Liquid Wrench can be applied to help free them.

Engine Oil

Every engine is engineered to run best with a specific "weight" of oil, so check the manual to see what's recommended, and use that. Oil weight is indicated by the numbers and letters on the bottle, which refer to **viscosity** — weight is the common term for viscosity — and quality. Viscosity is the measure of the thickness, or flow rate, of the oil. Lower numbers signify thinner oil; higher numbers mean thicker.

Check Plugs and Dipsticks

Most engines have a dipstick attached to the oil fill tube cap for checking the engine oil level, but some small engines instead use a **check plug**. Instead of checking the height of the oil film on a dipstick, you unscrew the check plug to see if oil flows out. If it doesn't, the level is low. Check plugs are also commonly used to monitor fluid levels in hydraulic systems and gearboxes.

Engine oil level is checked either with a dipstick or a check plug. Hydraulic fluid and gearbox oil are most often checked with a check plug.

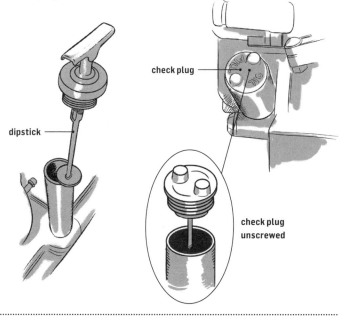

check plug

dipstick

check plug
unscrewed

Thin oils work well in low temperatures or under light loads; thicker oils are necessary for high temperatures and heavy loads. Small gas engines work at high temperatures and full loads and generally require oil in the SAE 30 range (SAE stands for Society of Automotive Engineers, which sets viscosity standards). Cars, tractors, and other machines with large engines that are operated through a wide range of loads and temperatures do better with multiple-viscosity oil. This type of oil has two numbers on the bottle, such as 10W–30, and contains additives that make it thin at lower temperatures and thicker at higher temperatures. In this example, the W means this oil is appropriate for winter use, the 10 is its viscosity when cold, and the 30 its viscosity when the oil is warm.

Oil quality, or **service rating**, is indicated by the letters on the bottle. This rates the engine-protecting quality of the oil, indicating the presence of additives to prevent rust and corrosion; reduce heat, soot, and carbon left over from combustion; seal compression; and dampen vibration. All categories but SH

Gaskets

Because fluids can seep through small seams between parts, gaskets — rubber, paper, cork, or soft metal — are often used in fuel, oil, and hydraulic systems to tightly seal connections between parts. Gaskets are replaced if they're leaking, or if they're exposed by disassembly.

Dealing with Used or Spilled Fluids

Oil, fuel, coolant, and other fluids can be a hazard to plants, animals, and groundwater if allowed to seep into the ground or poured down a drain. Always put old fluids in covered buckets for proper disposal. Old oil and hydraulic fluid can be dropped off at a hazardous fluids collection site in your area, or you may find an auto or engine repair shop with an oil-burning heater that will be happy to take the stuff off your hands. Antifreeze (sweet tasting and poisonous to animals and children) and other fluids should be turned in at hazardous waste collection sites.

Spills (and we all have them) can be covered with kitty litter or shop rags or even dirt to absorb the liquid, then scooped up and bagged until it can be properly disposed of.

and SJ are now considered obsolete; either of those two is fine for farm engines. A C instead of an S means the oil was formulated for diesel engines.

Finally, synthetic oil can be substituted for the same weight of petroleum-based oil. Though it's more expensive, it may make engines easier to start in the winter and retain its protective and lubricating qualities for a longer time.

FUEL

THE OCTANE RATING of gasoline should be appropriate for the engine. In cars, a high-octane gas allows for higher compression before combustion, delivering more power and less of the preignition that causes engine knocking. In a small engine, though, many mechanics say that a high-octane fuel will burn the valves and pistons and ruin the spark plugs. Fuels containing ethanol or alcohol also burn hotter and may have the same effect.

On the other hand, some mechanics and owners say that high-octane fuel gives better performance. See what your manual recommends for a particular engine, or try both types

....

Filters

Fluids must be free of contaminants for optimal engine performance. Small engines have air filters and fuel screens for this purpose; large engines have air and fuel filters, plus oil and hydraulic filters. These should be cleaned or changed as directed in the owner's manual.

AIR FILTERS COME IN ALL SIZES and shapes to fit the variety of engines found on most small farms. They're cheap and simple to replace on small engines; on tractors, you may have to disassemble and clean them once a year.

....

of fuel to see what works best for your equipment. We use 87-octane regular gasoline with 10 percent ethanol.

In old tractors with engines built to use leaded gasoline, *add lead replacer each time you gas up.* This costs just a few dollars a bottle at any farm store and protects the valve seats from rapid wear. Adding more than the recommended amount per gallon won't hurt, so don't skimp. If your tractor was built after 1975 or has had an engine rebuild, it probably has hardened valve parts, so lead replacer isn't necessary.

Diesel

Larger tractors generally are designed to use diesel fuel instead of gas. Diesel will thicken and gel at low temperatures; if you plan to use that engine during the winter, use a winter diesel blend if available, and store the machine in a heated shed, or install a plug-in block heater on the engine.

Winterizing Fuel

When storing gas engines for long periods, add a **gas stabilizer** to the tank. This prevents gumminess, which can make the needle valve on the carburetor float bowl stick so the engine won't start. After you've added stabilizer (we use Sta-Bil), run the engine for 5 to 10 minutes so the additive has time to work its way through the carburetor. Then top off the gas tank; this keeps water from condensing in the tank and fuel line as the engine cools, another frequent cause of engine problems.

AIR

WITHOUT AIR, fuel will not ignite; up to 9,000 gallons of air are needed to burn a gallon of gas. If the air is full of dust and grit, this will scratch and abrade the combustion chambers and work its way into the rest of the engine to cause more wear. For this reason, air intakes on all engines are equipped with some type of filter. Air also carries away the by-products of combustion through the exhaust system.

COOLANT

BIGGER ENGINES, like those on tractors, need more cooling than is provided by air movement, and so utilize liquid cooling systems. A mix of one-half antifreeze and one-half distilled water is the standard coolant formula; to maintain the right ratio you can premix the two and store in labeled jugs. **Coolant** is circulated around the engine to absorb and carry off excess heat to the finned radiator.

Using tap water rather than distilled water is okay, but you'll get more deposits and corrosion in the radiator core. If you dilute the antifreeze with too much water, the coolant is more likely to freeze in cold weather and boil over at high temperatures. Lastly, never remove a radiator cap until the engine has cooled off, and don't add cold coolant to an overheated engine, since this can crack the block.

Note that antifreeze is highly toxic and sweet tasting to animals; spills should be immediately wiped up, or, if your shed has a dirt floor, scoop up the contaminated dirt and dispose of it where no animal can find it.

BRAKE, POWER STEERING, TRANSMISSION, AND HYDRAULIC FLUID

GEARBOXES ARE LUBRICATED by heavyweight transmission oil, and no substitutes should be made for the weight specified by the owner's manual. Large engines may also use power steering fluid and brake fluid to operate those mechanisms, and though engine oil may be substituted in an emergency, the system should be repaired and refilled with the correct fluid as soon as possible.

A hydraulic system works by pushing noncompressible liquid through hoses, fittings, and hydraulic cylinders to lift or swing an implement or attachment. Use only hydraulic fluid in a hydraulic system, although in cold weather in small systems (such as a hydraulic wood splitter), lighter automatic transmission fluid can be substituted, for easier starting.

ADDITIVES AND AIDS

FLUID AIDS FOR ENGINES INCLUDE lead replacer for old tractors, fuel additives to clean carburetors, and flushes for radiators and **crankcases** (the lower part of the engine block where the crankshaft is located). Another type of additive promises to plug radiator leaks; this is not recommended by most manuals since it may also clog the smallest channels in the radiator, and it is only a temporary fix.

Aids for non-engine parts include penetrating oils to free frozen bolts; adhesive coatings (such as Loctite) to keep bolts from vibrating loose; and degreasing solvents such as mineral spirits, kerosene, and the like (don't use gas — it's too flammable), for cleaning such things as seed tubes, and to remove gummy deposits. Some spray solvents for carburetor and brake drum cleaning have more aggressive solvents, such as toluene or acetone, which can be hard on plastics and paints.

....

Keeping Track

Even small farm operations can wind up with many engines and implements, each with its own capacities and requirements for oil, fuel, and other fluids, as well as belt sizes and tire pressures. Listing this information on a single page and hanging it next to the tool bench is much quicker than paging through owner's manuals for the information whenever you want to change oil or do other maintenance. Also helpful is keeping a diary of what maintenance and/or repairs were done on what date to which machine (see page 103).

....

CHAPTER FOUR

TOOLS

Although any farmer worth the dirt under his or her fingernails is creative about making emergency field repairs and jury-rigging something without the right tools or the right parts, there are some basic tools for equipment you should have in your shop, along with a few specialty tools for specific jobs that will pay for themselves in time and effort saved.

We've found over the years that two sets of the small hand tools saves some discord over who left what where. We've also found it's important to take the time to return tools to where they're supposed to be. A traveling set of tools for field repairs is handy as well.

THE BASIC TOOL KIT

BELOW IS A BASIC LIST of tools you should have for equipment maintenance and repair, followed by a discussion of those tools that may not be self-explanatory. You may find other tools that might be useful in your shop by spending some time in the tool section of the farm store on a rainy day.

General Purpose Tools

- Hammers
- Short-handled maul (about 3 pounds)
- Screwdrivers: several sizes of both flat and Phillips heads. A magnetized screwdriver is handy for small screws and bolts.
- Pry bars: small, medium, and large
- Punch set: for pushing out pins and aligning holes
- Pliers: Standard, electrician's, needlenose, and locking
- Hacksaw with metal-cutting blade
- Pipe wrenches: a pair of them allows you to hold a pipe on both sides of a joint
- Lug bolt wrench, automobile type
- Bench vise: for holding parts while you sharpen, adjust, rivet, or do other small-scale work
- Propane torch (small) for heating frozen bolts to loosen them
- Large and toothbrush-size wire brushes for cleaning parts, bolt threads, and so on
- Shop magnet for finding lost small parts

PLIERS

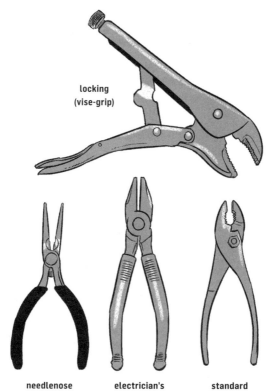

locking
(vise-grip)

needlenose   electrician's   standard

Power Tools
- Drill and selection of bits for metal
- Impact wrench for quick bolt removal and installation
- Grinder: fixed or hand-held angle
- Air compressor to fill an air storage tank
- Shop vacuum
- Power washer

Basic Tools for Working on Equipment
- Owner's manuals
- Grease gun, preferably with flexible nozzle
- Sets of box-end and open-end wrenches
- Socket wrenches: several sizes of driver; set of metric sockets; set of American sockets in both shallow and deep lengths; extensions and adapters
- Breaker bar (a.k.a. breakdown bar), a long-handled socket driver
- Adjustable (Crescent) wrenches, in a couple of sizes
- Set of Allen wrenches (a.k.a. hex-head wrenches); these are available in both American and metric
- Spark plug socket (for socket wrenches); several sizes may be needed, depending on differences between engines
- Bolt cutters
- Nut splitter, a sort of screw-down chisel for breaking off rusted-on nuts
- Air gauge and portable compressed air tank (for filling tires)
- Jacks and blocks
- Jumper cables
- Battery charger

Specialty Tools for Specific Jobs
- Oil filter wrenches, small and large
- Gapping tool or set of gappers for spark plugs
- Spring clip tool (a.k.a. snap ring pliers), reversible so it will contract and expand spring clips
- Bearing puller (expensive!)
- Riveting tool
- Voltmeter
- Welder, if you have the skills

Supplies
- Tubes of grease
- Lubricating oil
- Dry lubricant
- Engine oil (correct weights for your engines)
- Antifreeze and distilled water
- Wire, bungee cords, rope, nylon cable ties in several sizes
- Penetrating oil
- Loctite or similar product
- Teflon tape for wrapping screw-together hydraulic fittings (to prevent leaks)
- Electrical tape for worn insulation on wires
- Electrical wire and connector kit
- Duct tape for general use
- Shop rags (heavy-duty paper towels)
- Set of funnels
- Nonflammable solvent, plus a bucket and brush for cleaning small parts
- Kitty litter for absorbing spills

MANUALS

THE MOST IMPORTANT TOOL to have for each piece of equipment is the owner's manual. This explains maintenance procedures and schedules, where zerks are located, how to make adjustments, and how to clean or change filters; a "troubleshooting" section, usually in the back, lists the causes of common problems with that particular machine. The list of capacities and specifications for fluids and tires is essential for proper maintenance.

If you buy used equipment and it doesn't come with a manual, you can in most cases find one on the Internet. Note also that many small engines now come with separate manuals for the engine and the machinery, and old tractors may have both an operator's manual and a maintenance and repair manual. In those cases, you'll want both manuals.

GREASE GUN

THE SECOND MOST IMPORTANT TOOL to own is a grease gun. This works similarly to a caulking gun; at the end of the tube (and we highly recommend getting a gun with a flexible rubber rather than jointed or straight metal tube) is a metal coupler that fits over a grease zerk. As you force grease through the gun's tube, the pressure depresses the spring bearing on the zerk, allowing grease to flow into the appropriate place on the equipment.

WRENCHES

WRENCHES ARE THE MOST USED TOOL in the box and are essential for dealing with bolts. There are several types. Open-end wrenches, box-end wrenches, and socket wrenches all come in sets, with one wrench or socket for each size of bolt head. Simple open-end and box-end wrenches take up less space in the toolbox but require you to reset the wrench on each turn when there's not enough room for the handle to make a complete revolution around the bolt head. Ratcheting box end wrenches solve this problem.

Socket wrenches have separate drivers (handles) and sockets (heads) with a ratchet mechanism so the wrench does not have to be removed and reset when turning a bolt head. A set may come with two or three different-size drivers and a broad selection of sockets so the wrench can be matched precisely to the bolt. Get some extensions, as well, to keep your knuckles away from the work. Sockets are more expensive and cumbersome than other wrenches, but faster and easier to use, especially in tight quarters. Sockets come in either metric or US sizes and both shallow and deep lengths; what you'll need depends on what size bolts are on your equipment.

A breaker bar (or breakdown bar) with a receiver for a socket is handy when you need extra force for stubborn bolts, or you can slip a piece of hollow pipe over a regular driver to get extra leverage. (This can also break the driver, so be careful.)

TYPES OF WRENCHES

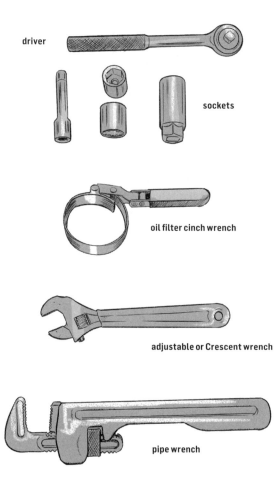

driver

sockets

oil filter cinch wrench

adjustable or Crescent wrench

pipe wrench

Other Wrenches

Adjustable (Crescent) wrenches have screw mechanisms in the heads to adjust the size of the gripping part of the tool and are good for field situations, since they replace an entire set of open-end, box, or socket wrenches. There are also several new adjustable designs on the market that look good. Continually using an adjustable wrench on a bolt may round the corners on the bolt head, eventually making it impossible to turn with any wrench.

Pipe wrenches turn rounded parts, such as pipes and shafts. **Allen wrenches**, also called key wrenches, come in sets and fit into the hexagonally shaped recesses on hex-head bolts. **Oil-filter wrenches** are the cinch or pinch type and don't dent filters. All three should be in your shop.

SPARK PLUG TOOLS

THE CERAMIC ON SPARK PLUGS is prone to cracking, so plugs are best installed and removed with a special padded **spark plug socket** that fits on a socket wrench driver. A second tool is needed to calibrate the gap between the plug body and the wire at the end and is called, logically enough, a **gapping tool**. This set of thin pieces of metal, each machined to a precise thickness, allows you to select the piece that matches the specification in the manual, insert it in the gap, and gently tap the wire to correct the gap.

TOOLS FOR WHEELS

LOW TEMPERATURES AND SLOW LEAKS will cause equipment tires to lose pressure. A good-quality air gauge makes it simple to check whether tires are inflated to the specifications in the manual. An air compressor and portable air tank (which is filled with air by the compressor) give you the convenience of being able to inflate tires on the farm or in the field, instead of having to load equipment on a trailer and run into town.

When a tire has to be removed for repair or replacement, the lug bolts can be taken off with a socket wrench, but a **lug bolt wrench**, a metal cross with different-size heads at each end, gives better grip and more torque.

JACKS AND BLOCKS

SMALL HYDRAULIC JACKS are compact and powerful, and good for raising equipment a small distance — bigger distances if you stop to add blocks under the jack as you go. They wear out after a few years. A large manual jack should last forever and can raise larger things a greater distance. *Both types can shift or tip.* We keep one of each.

Jacks should not be relied on to hold a piece of equipment off the ground for any length of time: Chock and block! Keep on hand some broad, thick, long blocks of wood, and stack them next to the jack once the equipment has been raised. Then reverse the jack and let the equipment settle the inch or so onto the blocks. It's also a good idea to place a broad, long board under the jack itself; this will help stabilize it and keep it from sinking into the dirt.

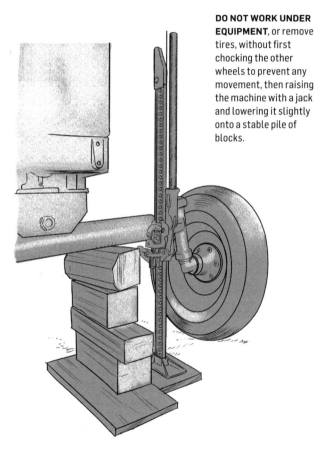

DO NOT WORK UNDER EQUIPMENT, or remove tires, without first chocking the other wheels to prevent any movement, then raising the machine with a jack and lowering it slightly onto a stable pile of blocks.

POWER TOOLS

MANY VARIANTS OF PLUG-IN and battery-powered wrenches, drills, drivers, saws, and grinders are available at a reasonable price. Most useful around machinery are an impact wrench for bolts, a handheld grinder for sharpening mower blades and grinding off bolt heads, and a drill with a set of screwdriver and metal bits. Plug-ins offer more power for the money; battery-powered tools are convenient when you're away from buildings.

Pneumatic power tools that run off an air compressor are also widely available, more powerful, more expensive, and generally beyond the needs of the average farm shop.

Cleaning Tools

A shop vacuum is invaluable for blowing or vacuuming dirt and debris off equipment, workbenches, and other surfaces. A power washer (pressure washer) is much more effective than a garden hose with a sprayer handle and has many other uses

....

Acquiring Tools

Tools come in three general categories of quality: cheap, good, and excellent. The quality of a tool is indicated by brand name, price, fit, finish, and the precision of any mechanisms. Cheap tools are not worth the price, and professional-quality tools may be overbuilt for the frequency and stress of use they'd have on the average small farm. A midrange tool is usually more than adequate, but when in doubt, buy the higher-quality tool.

....

around the farm besides cleaning equipment. Never power wash any part, such as a bale chamber or seed bin, that must be dry and rust-free to function properly.

SPECIALTY TOOLS

MANY SPECIALTY TOOLS are available to make mechanical work easier and more precise. Before buying anything costly with a limited use, consider whether you will use it often enough to justify the price. Many of these types of tools can be rented or borrowed (be sure to return them promptly and in good condition if you want to stay on good terms with your neighbors).

There are also a number of affordable specialty tools that will pay for themselves in frustration saved, even if they are only rarely used.

An inexpensive **spring clip tool**, a specialized type of pliers for removing spring clips and which can be reversed so it will both compress and expand a clip, saves a lot of time and lost spring clips. It's difficult to remove or reinstall chain links with anything besides a **chain-link tool**. A **bearing puller** for extracting stuck bearings from wheels, pulleys, or sprockets is fairly expensive, but like the others, it easily does a job that can be difficult and frustrating when attempted with more general-purpose tools.

A **voltmeter** measures the charge in a battery and is also used to track down elusive electrical shorts. Voltmeters range from cheap, simple tools to expensive, multifunction units with their own manuals.

CHAPTER FIVE

MAINTENANCE

Regular maintenance (we've said it before and we'll say it again) is the single most important thing you can do to minimize breakdowns and repairs. Maintenance means cleaning, inspecting, lubricating, changing fluids and filters, and adjusting equipment on a regular schedule for optimal performance, long life, and minimal breakdowns.

Whether you are using, maintaining, or repairing equipment, *safety must be your top priority*. Equipment can maim or kill you if you're in the wrong place at the wrong time: Working on equipment puts you in a position to be hurt by equipment. Turn off the engine and PTO, disconnect spark plugs, block and chock, and use your common sense.

GENERAL MAINTENANCE

THE FIRST THREE STEPS in a good maintenance program are:

1. **Read the owner's manual.** This specifies what to do and how often, and explains how to adjust that equipment for best operation.

2. **Store equipment inside.** This minimizes rust, faded and brittle paint, and the accumulation of debris and dirt. If you don't have a shed, strap a tarp over delicate mechanisms, such as the knotter on a square baler.

3. **Keep equipment reasonably clean.** You don't have to be able to eat off the tractor seat, but blowing off or power washing on a regular basis keeps dirt out of mechanisms and makes it easier to see what you're doing when you're working on the machine.

Remember that *the most important maintenance you can do to keep equipment running is regularly greasing all zerks and keeping engine oil clean and the reservoir full.* Check the owner's manual for the location of zerks, since they can be difficult to see; replace any that have broken off or fallen out. Also, wipe a light coating of grease on shafts in sleeves, connecting parts, and sliding parts that don't have zerks.

Maintenance Schedule

Maintenance should be done on a regular schedule, but not everything needs to be done every time you start an engine or pull out an implement. The list below describes how often to do specific maintenance procedures. ***Starred items** are tasks that require more explanation and are discussed on the following pages.

ENGINES
With Every Use
- Check oil* and fuel levels; add more if indicated.
- Clear radiator and air intake screens if dirty with a vacuum or soft brush. (On chain saws, remove the back cover to get at the screen; the wood fiber that collects there can be brushed off with an old toothbrush.)

Several Times a Season
- Check coolant level (remove cap, and see that the coolant covers the radiator core); add more if low.
- Check outside of radiator (on large engines) and *gently* clean if dirty; an old soft paintbrush is good for this job.

At Least Annually
- Change engine oil and filter on all four-cycle engines.*
- Clean precleaner and oil bath air filter on tractors according to the directions in the owner's manual (do this at the same time you change engine oil; it should be the same type of oil).
- Change or clean air filters on small engines.*
- Inspect coolant; flush system if indicated.*

- Inspect spark plugs; adjust, clean, or replace if indicated.*
- Check manual for other annual tasks specific to that engine, such as cleaning the crankcase breather element (this is an air pressure equalization device found on some four-cycle engines).

As Needed
- Check fuel filter on small engines and change if indicated; see the owner's manual for directions.
- Clean the fuel strainer and sediment bowl on tractors according to the directions in the owner's manual.

EQUIPMENT
With Every Use
- Visually inspect tires for proper inflation; if a tire looks low, use the air gauge to test pressure (the correct pounds per square inch — psi — will be on the side of the tire). Use a portable air tank to reinflate the tire.
- Visually inspect moving parts for function, tightness, and proper adjustment; tighten and adjust as indicated.*

Several Times a Season
- Grease zerks.
- Oil chains.
- Sharpen blades.*
- Visually inspect metal frame and bolts, rivets, springs, and so on for unusual wear (look for shiny spots on the metal or round holes becoming oblong, for instance) or cracks. Tighten, replace, or repair, as indicated.

- Inspect discs, sweeps, tines, teeth, and so forth for looseness, wear, and proper alignment, and adjust or replace as indicated.
- Check hydraulic fluid level; add fluid if indicated.
- Check hydraulic system for leaks; tighten connections or replace couplings if necessary.

At Least Annually
- Clean by blowing off with air or rinsing with water.
- Visually inspect chains for tightness; tighten if indicated.*
- Visually inspect belts for fraying and stretching; replace if indicated.*
- Apply spray-on belt dressing to all belts.
- Check condition of hydraulic fluid; change fluid and filter if it's milky or dirty.*
- Check condition of hydraulic hoses, and replace if worn.*
- Put a light coat of lubricating oil on bolts and all bare metal parts, except brakes, clutches, and the interior of metal planter boxes and seed tubes. On these last two, use a dry lubricant or light coat of spray-on metal paint (since these won't collect dirt) as a rust preventive.

As Needed
- Review manual for inspections, adjustments,* and other maintenance specific to the implement.
- Repack wheel bearings if the wheel is off for some other reason, or if bearings have been frequently submerged in water or mud.*

Visual Inspection

Taking a few minutes before or after using a piece of equipment can prevent small developing problems from turning into major headaches. Make sure chains and belts are appropriately tight and not showing signs of excessive wear. Look for loose or missing bolts, clips, and pins, and for unusual wear patterns — often indicated by shiny metal, or holes (such as on hitches) worn out of round. Inspect also for leaking engine oil, hydraulic fluid, gearbox oil, coolant, and fuel. Note if tires are low or wobbling, if debris is clogging the pre-screen on the air filter, and anything else that doesn't look quite right.

Changing Engine Oil

Every time you gas up a four-cycle engine, check the engine oil level and condition. Do this when the engine is not running: Pull out the dipstick, wipe it clean with a paper towel or rag, reinsert it, and pull it out again to get an accurate reading. If the oil film is below the "fill" line on the dipstick, add oil (using the weight specified for that engine by the owner's manual).

Do not overfill; this causes foaming of the oil and decreases engine efficiency and life span. If the oil is dark and opaque instead of light and clear, or milky (indicating water in the oil), it needs changing. If the engine has an oil filter, change this at the same time. Changing oil takes about 10 minutes.

How to Change Oil

1. Run the engine long enough to get it thoroughly warm, park it on level ground, and turn it off. On a small engine, disconnect the spark plug wire for safety.

2. Put a bucket or basin under the oil drain plug (some engines may have a tube).

3. Unscrew the plug, and drain the oil into the bucket.

4. If there is a filter, remove it with an oil filter wrench, and replace it with a new one, first wiping the bottom of the new filter or its gasket (if it has one) lightly with oil to ensure a good seal. Snug the new filter down by hand, and give it an extra quarter turn with the wrench.

5. Replace the drain plug; snug it down firmly, but don't overtighten.

6. Add the specified amount and type of new oil at the oil cap or the dipstick tube.

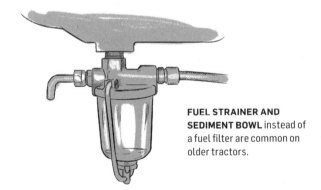

FUEL STRAINER AND SEDIMENT BOWL instead of a fuel filter are common on older tractors.

Frozen Bolts or Loose Bolts

A rusted-in bolt is one of the most common frustrations you'll deal with on old implements, since much maintenance and most repairs begin with removing bolts to get at a part or system.

Start by squirting on some penetrating oil (Liquid Wrench is one brand) and leaving it for a couple of minutes up to a couple of days. If that doesn't work, give the bolt head or side of the nut a couple of light taps with a hammer — not the shaft end; that can deform the threads and make the problem worse. If it's a hex-head bolt, you can put the socket on a breakdown (a.k.a. breaker) bar for more torque or put a piece of pipe over the regular wrench handle and give it some controlled pulls (a power impact wrench can be very useful here).

If it still won't come loose, try heating the bolt or nut with a handheld propane torch to expand the metal and break the seal. If all else fails and the bolt is held by a nut, put a grinding wheel on your power drill, and grind the head or the nut off the bolt so it can fall out of the hole. If a frozen bolt is threaded into a receptacle, the job becomes more difficult. Machinists and mechanics might weld a bar or fresh-edged nut onto the rounded bolt head to screw it out.

Bolts that often vibrate loose can be helped to stay in place by coating the clean, dry threads with an adhesive product, such as Loctite. Other things to try are adding a split washer, a locknut, or a second jam nut on top of the first nut. Keep a variety of spare bolts handy for emergency replacement of broken or lost bolts.

Air Filters

Small and newer engines will have an air filter of paper, foam, metal, or some combination. Clean or replace these according to the directions in the manual, or at least annually — it's a quick, simple job. Paper filters can sometimes be shaken or vacuumed and reused but should be replaced when they get really dirty.

Radiator Flushing

If coolant appears dirty or milky, flush the cooling system.

How to Flush a Radiator

1. Open the drain plug at the bottom of the radiator, and drain the coolant into a bucket. (Cover and label the bucket; dispose of at a hazardous waste collection site.)

2. Close the plug, and fill the system with water or a radiator flush. Run the engine for a few minutes, then drain the flush into a labeled bucket.

3. Refill with new coolant.

Spark Plugs

To remove a spark plug, gently wiggle off the spark plug wire, then use a special padded spark plug socket on the socket wrench to unscrew the plug; this prevents damage to the plug's porcelain. If the small L-shaped bare wire at the top of the plug has a brown coating (indicating good ignition), gently clean it off with a wire brush. If the coating is black or wet, or the porcelain is cracked, the plug should be replaced with the same size and type as specified in the owner's manual. On engines with several spark plugs, *do not mix up the order of spark plug wires;* this will disable the engine.

To check the spark plug gap, consult the manual for the correct setting, then insert the same-size spacer from the gapping tool in the space between the wire and the body of the plug. To correct the gap, lightly tap or pry the wire until the gapper is just touching both wire and body.

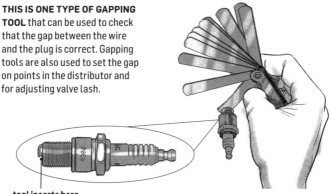

THIS IS ONE TYPE OF GAPPING TOOL that can be used to check that the gap between the wire and the plug is correct. Gapping tools are also used to set the gap on points in the distributor and for adjusting valve lash.

tool inserts here

Hydraulic Fluid and Filter

Any time you attach hydraulic couplings, wipe off the ends first to keep dirt out of the system.

Several times a season, visually inspect hoses and couplings for wear and leaks. Replace a worn hose by removing the hose and associated couplings and taking it to an equipment repair shop. They will cut a new hose and put on the old couplings for a reasonable price.

Leaking couplings should be disassembled and cleaned and the threads wrapped with Teflon tape. If this doesn't work, have the couplings replaced.

Since hydraulic fluid is both long-lasting and quite expensive, change it only if it becomes milky (indicating water in the system) or noticeably dirty, or there appears to be a partial clog in the system. This is done the same way you change engine oil: Run the engine so the fluid warms and holds as much dirt as possible in suspension; then drain it, change the filter, install a new filter, and refill the system with new fluid.

Hydrostatic drives and power steering units may be sealed and require no regular maintenance, or they may share fluid with the main hydraulic system. Hydraulic brakes generally have an accessible reservoir that can be checked for brake fluid. Consult your manual.

Sharpening Blades

Most blades on brush and lawn mowers are **chisel ground** — flat on one side and angled on the other. Other types of blades are mostly **flat ground**, where both sides are angled to the point. Blades can be removed for sharpening, or sharpened in place if there is room to get the right sharpening angle with your tools. Depending on the amount of metal to be removed, a file or sharpening stone may be sufficient. A grinding wheel, belt sander, or handheld power grinder (our preference) is quicker, but take care not to overheat the metal, which will change the tempering (hardness). *When using power grinders, wear eye and hand protection to avoid injury from sparks and cuts.*

Adjusting Implements

If all parts are lubricated and functioning but equipment is still not running evenly or easily, consult the owner's manual for instructions on how to adjust tensions and spacings for better performance. Often it will take a few tries to get everything working the way it should. To find out if changing adjustments has worked, run the equipment and see how it performs. If something isn't quite right, change the adjustments and run it again. An experienced neighbor can be very helpful in these situations.

Changing Belts

Since belts fray, stretch, and break fairly often, it's worthwhile to keep extras on hand. Label the spares (a silver-colored marker pen works well) with their size and length and what piece of equipment they belong to.

Instructions for replacing a belt are found in the owner's manual, or you can look for the tensioner pulley, spring, or guide that holds the belt tight; loosen this to get the old belt off and the new one on. If there is no tensioner, get the belt around the smaller pulley, then feed it into the larger one as you turn it slowly by hand till the belt flips over the lip into the groove.

Repacking Wheel Bearings

Sealed bearings require no maintenance; when they fail, they're replaced. Unsealed bearings should be cleaned and repacked, usually annually but less often if the equipment is used only occasionally. To repack bearings, remove the wheel, then remove the grease cap, spindle nut, and cotter pin to get at the bearings. Remove the bearings in their raceway, and use solvent to clean out the old grease. Dry the bearings and raceway, then carefully work in the fresh grease. If the bearings are stuck on the shaft, you can clean and repack them in place with bucket and brush or spray solvent.

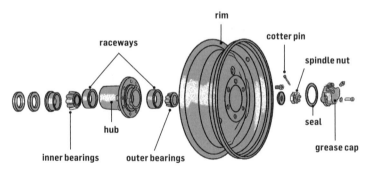

A STANDARD WHEEL AND BEARING assembly from an old John Deere tractor

Tightening and Repairing Chains

To tighten a chain that has stretched, first try adjusting any tensioners, if present. If this doesn't work, try removing a link and replacing it with a half link. Broken chains can be repaired by replacing the bad link. Chain-link kits, complete with instructions, are available cheaply at farm and hardware stores. Take a link along to the store to make sure you buy the right size repair kit.

PREPARING FOR COLD WEATHER

AN EXCELLENT TIME TO do annual maintenance chores is before you store equipment for the season; this helps ensure quick and easy start-up in the spring. Don't forget to drain the water line in the pressure washer and add RV (recreational vehicle) antifreeze; or simply move the unit into a garage or basement where the temperature won't fall below freezing.

When storing an engine for the season, add a gas stabilizer to the tank, then run the engine for several minutes to make sure the additive gets to the carburetor, where it will keep the needle valve from gumming up. Then top off the gas tank to prevent water condensation in the fuel line. Gumminess and water in the fuel system make engines hard or impossible to start.

Many manuals recommend that before storage you mist the engine cylinders with lubricating oil to prevent rust. (Do this after you've finished running the stabilizer through the carburetor.) This is done by removing the spark plug(s), spraying or squirting a tablespoon of engine oil into the cylinder, then cranking the engine two to three times without turning it on, so the pistons move and coat the inside of the cylinders with the oil.

Equipment that's not stored but used during cold weather often runs more easily with a few adaptations. Put automatic transmission fluid in a wood splitter's hydraulic system instead of hydraulic fluid; it will be much easier to start with the pull cord. Get a snowblower with a plug-in electric starter for the same reason. Chain saws don't come with plug-in starters, but they are small enough to bring into the relative warmth of the basement overnight. Tractors can be fitted with plug-in engine heaters; plug in your tractor on a cold night if you plan to use it the next day.

CHAPTER SIX

DIAGNOSTICS

The cause of a malfunction or breakdown may be obvious — a flat tire, for instance — or obscure, and often with several possible causes, as when there's an elusive short in the electrical system. If you can see what's wrong, skip to chapter 7 on repairs; if you can't, start by consulting the troubleshooting section in the back of your owner's manual for problems specific to a particular implement or engine. If, as is often the case, the manual doesn't help, then identify which system appears to be the source of the malfunction, and examine it part by part to pinpoint the malfunction or consider the suggestions outlined in this chapter.

If you work your way through the system where the problem seems to lie and don't find anything, then look at the other systems. Sometimes what seems to be, for example, an electrical malfunction turns out to be caused by something completely different, like a dead bug partially clogging the fuel line. If you still can't find the source of the malfunction, or you do find it and realize it's beyond your skills to repair, then it's time to get help from a knowledgeable neighbor, friend, relative, or professional mechanic, as will be discussed in chapter 8.

Note that catching a malfunction early rather than late tends to keep a little problem from turning into a bigger one. Abnormal noise coming from an engine or implement is very often your first clue that something isn't right. If running equipment doesn't sound, smell, or look right, stop it immediately, and look for the source of the problem.

START WITH THE SIMPLE AND OBVIOUS

SOME OF THE MOST COMMON causes of problems are:

- **Operator error.** Read the operating instructions to learn what a piece of equipment is and is not capable of doing, and under what conditions.
- **Inadequate lubrication.** Check the oil level, and add more if needed. Make sure zerks are greased and chains oiled.
- **Dirt and debris in systems and mechanisms.** Make a visual inspection to see if anything is clogged or plugged.
- **Parts not properly adjusted or aligned.** Make a *safe* (that is, from a distance) visual inspection of moving parts while the engine is running, and a visual inspection of working

parts when the implement is engaged. Watch for stuttering, grinding, excessive motion, and similar indications that something isn't working smoothly.

- **Parts worn beyond the point of being functional.** Visually inspect all parts, especially belts, chains, and blades. If you're not sure how much wear there is, compare it to a new part.

When the cause of a breakdown is not immediately obvious, always start by checking the basics.

Engine

An engine that won't start, or starts but runs roughly, or lacks power, is most commonly not getting enough of one of the big three: fuel, spark, or air.

Fuel

- Check if there's gas in the tank. (Water in the gas also causes problems; suspect this if the engine has been in long-term storage.)
- Check that the choke is in the correct "start" or "run" position.
- If you can smell raw gas, the engine is probably flooded. Leave it for a few minutes, then try to start it again, with less choke.
- Check that the fuel line in a gravity-fed system is clear by closing the fuel line valve, then detaching the line at the carburetor end, and inspecting the line and the screen. Open the valve, and make sure fuel flows through the line. (Engines with an electric fuel pump require battery power to move fuel.)

Spark

- The battery might be dead. If you can jump-start the engine, this is the problem.
- A spark plug wire may be loose or has fallen off.
- If the engine has been exposed to splashing water, there may be water in the distributor cap.

Air

- The choke may be stuck. Unless you can look down the air intake to see the valve (not too likely), you have to rely on making sure the cable that goes to the carburetor is moving when you move the choke. If it does move, then the problem may be inside the carburetor.
- There may be debris or a mouse nest in the air intake/exhaust system. (Mice can rapidly make nests or food caches in stored equipment; they also will chew wires.)

Connections

- Check if a hose has come loose or broken.
- Check if the hookup is correct (hitches and hydraulics).
- Check that chains and belts are in place, aligned, and appropriately tight.

Other Parts

- Check for dirt, crop, or debris plugging parts or systems.
- Visually inspect for broken or missing parts.

  If these obvious potential causes check out okay, then:

  1. Identify the system or mechanism that appears to be involved.

2. Work your way through the components of the suspect system or mechanism until you find a possible source for the problem.

3. Fix it, run the equipment to see if that was indeed the problem, and if not, repeat the first two steps.

There is one caution here: Most everyone has fallen into the trap of replacing one part after another until the problem is found. If the parts are cheap and easy to install, this is not a big deal. But if you go to the time and expense of replacing something such as a starter motor when the problem is nothing but a corroded battery cable, you'll be upset. Always start diagnostics by thinking through the problem, and start blind repairs with the simple, cheap, easy stuff. Below we'll discuss how to inspect and test the parts of a system, beginning with the five that make an engine run.

ENGINE PROBLEMS

ENGINE PROBLEMS THAT ORIGINATE inside the engine block are beyond the reach of this beginner's book, since they can be difficult to diagnose, hard to access, and fussy to repair. For these, you're usually better off hiring a professional than tackling it yourself, unless you have the time, tools, and expertise. Engine problems that originate outside the block are caused by a problem in one of the previously discussed (see chapter 2) systems — electrical, fuel, lubrication, cooling, and air intake/exhaust.

Problems in the Electrical System

Some common symptoms of a problem in the electrical system are:

- No noise when you turn the ignition
- A clicking or whining noise when you turn the ignition
- A reluctant, slow turnover of the engine when you turn the ignition
- An engine that turns over but won't start
- An engine that starts but lacks power and makes a stuttering exhaust noise (lack of power can also be caused by problems in other systems)

Components in the electrical system to check (in order of appearance) include the battery, ignition switch, solenoid and starter motor, coil, distributor cap and rotor, spark plugs and wires, the alternator/generator, and all the wires that connect these parts, including any lights and gauges. In a non-gravity-fed system, the fuel pump is also run by electricity. Start with a visual inspection of these parts and then, if the problem still isn't apparent, use a multimeter to locate the fault.

Checking the battery. If the battery appears dead, it's a failed battery, a short somewhere in the system that is draining the charge when the engine is off, or a faulty re-charging system. Begin by inspecting for corrosion on the terminals, and for tight connections. Be sure also to check the *ground* end of the negative terminal where it attaches to the tractor frame. If these are okay and there's still no noise when you turn the ignition, the battery is dead. Charge the battery and then use a multimeter (or disconnect it and let it stand overnight) to see if the battery takes and holds the charge. If it doesn't, then the battery is faulty and should be replaced. If the battery does hold a charge, then a short in the electrical system is draining the charge when the engine is not running, or the recharging system is weak.

Checking spark plugs and wires. An engine that turns over but doesn't catch, or sputters, can indicate a problem at the plugs. Check the spark plug wire(s) to make sure they are not broken and are solidly attached to the top of the plug and to the distributor cap. Bad wires are easily replaced.

If the wires are okay, unscrew the plug(s) and inspect for cracks in the porcelain, black sooty deposits on the tip wire, or a gap that's out of adjustment. Adjust the gap if needed, take soot off with a wire brush, and replace cracked plugs.

If the plug(s) look good, reconnect the wire to the plug, touch the threads of the plug to a metal frame part, and crank the engine. If there's a spark, then current is present, the plug is good, and the problem lies elsewhere in the system.

....

Battery Care

Never let a battery freeze; that will ruin it. A fully charged battery won't freeze, so in cold weather, start engines occasionally and let them run to re-charge, or keep a trickle charger on the battery when it's in storage during the winter. If the battery isn't sealed (old batteries), check the fluid level before putting it into storage, and top up with distilled water if it's low (*never* add battery acid to a battery). Or you can do as the old-timers did: disconnect the battery and store it in the basement over the winter.

....

If there's no spark, make doubly sure the spark plug wires are good by testing them with an ohmmeter. If the wires are functional, then look for the source of the problem elsewhere in the system.

Checking the distributor cap and rotor. Engines with more than one spark plug have a distributor that delivers spark to each plug in succession. Visually inspect the cap for looseness or cracks, and make sure the coil-wire-to-cap connection is tight (this is the wire in the center of the cap). Remove the cap and inspect the rotor inside for dirt, moisture, breakage, or excessive wear on the contacts and points. Problems with these parts are fixed by replacing the part.

If cap and distributor parts look fine, then any engine "missing" may be due to spark delivery timing being off. This is checked with a timing light, which takes advanced or professional expertise.

Checking the ignition switch. Put a voltmeter between the input (wired to the battery) and output (wired to the coil and starter) terminals at the back of the switch. If voltage is flowing to the switch, through the switch, and away from the switch when the switch is on, then it is working.

Checking the coil. The coil is a transformer that steps up voltage to the distributor. If there's no spark at the plugs or the distributor, and the wiring, battery, and ignition switch have checked out, then a faulty coil is the likely culprit.

Checking the solenoid. A starter whine without engagement of the starter motor is indicative of solenoid malfunction. Solenoids are generally replaced rather than repaired.

Using a Multimeter

Multimeters (combined voltmeter, ammeter, and ohmmeter) combine several ways of measuring electricity in one device. The **voltmeter** function measures the voltage difference between two points in a charged system. If there is a point of resistance between the measuring points, as in a component where work is extracted, there should be a voltage drop. The **ohmmeter** function measures resistance (ohms) in an uncharged circuit by sending out and receiving its own test charge. The **ammeter** function measures the actual flow (amperage) of current in a charged system.

A multimeter is used to test for three common problems in an engine:

- **Dead battery.** After charging the battery, use the voltmeter function to test if the battery can achieve and hold its listed charge (6 or 12 volts). If it doesn't, the battery has failed and should be replaced.

- **Wire break or bad connection.** If there's no visible problem, test wire section by section with the ohmmeter function to pinpoint breaks, corrosion, or loose connections. High resistance and pent-up voltage indicate a wire is not conducting electricity efficiently between two points.

- **Short in the wiring.** Current always follows the path of least resistance to complete a circuit. When a loss of insulation allows a bare wire to touch the frame, this allows the positive flow of current to go directly to the negative (neutral) ground without having to flow through the rest of the circuit or wait for a switch to be turned on downstream. Use the ammeter function to see if amperage is flowing when the destination component is not receiving current.

Checking the starter motor. A clicking noise without whining or grinding indicates the solenoid is engaging, but the starter motor is not turning. If the battery and connections are good, then the motor has failed and will need to be replaced.

Checking wiring. Tracing a break or short in wiring with a multimeter can be tedious and frustrating when there's no visually obvious problem. If you can't find a slow, draining short or don't have the time, you might be able to install a cutoff switch on the terminal wire between the battery and the ignition switch so that a short won't drain the battery when the engine is off.

IF A BATTERY IS ALWAYS DEAD when you want to start the engine, you can use a voltmeter to test whether it is still capable of holding a charge. If it is, the problem is elsewhere in the system.

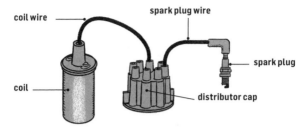

ON A BIG ENGINE, voltage is amplified in the coil and sent to the distributor via the coil wire, which plugs into the middle of the distributor cap. The distributor then sends a timed charge to each of the spark plugs in turn, via the spark plug wires.

Problems in the Fuel System

If the malfunction appears to be in the fuel system, but there is fresh, water-free gas in the tank and no blockage in fuel line and filter, then look at the choke, throttle cable, carburetor, and idle adjustment. If there is a fuel leak, wipe the area dry so you can see exactly whether it's a clamp, gasket, line, gas tank, or other part that is leaking.

Checking choke function. An engine that won't start when cold, or starts and runs roughly, may indicate a choke stuck in the open position. A choke that is stuck closed will cause engine flooding, and you'll smell gas. Check that the external controls (usually a switch and cable that runs to the carburetor) move freely. To see if the choke valve is operating freely, look down the air intake to the carburetor while you work the choke (if possible).

Checking the throttle cable. This requires consulting the owner's manual to determine what the correct adjustment is, and how to make it.

Checking the idle adjustment. As with throttle adjustment, this will be a little different for each engine, so it's best to follow directions in the owner's manual.

Checking carburetor function. An engine that won't start, or lacks power, and/or emits dark exhaust smelling of gas can indicate that the float valve is stuck open inside the carburetor. A float valve that is stuck closed, or a gummed-up Venturi jet (atomizer) can also keep an engine from starting, or cause a surging idle, and chugging or running roughly on acceleration. If you tap on the carburetor and the noise changes, it's likely that something in the carburetor is sticking, or there is debris in the bowl.

Problems in the Lubrication System

Lubrication system malfunctions that aren't due to low engine oil levels or a plugged filter indicate a possible problem inside the engine block that will need professional attention. A milky appearance to the oil means water in the system, which can indicate a head gasket leak, while white exhaust smoke indicates oil in the combustion chamber caused by failed rings or valve stem wear. A gauge that indicates low oil pressure despite adequate oil level can mean excessive parts wear, a debris blockage, or a failed oil pump.

Problems with the Air Intake/Exhaust System

A lack of engine power, starting then sputtering and quitting, and/or black exhaust smoke can be symptoms of problems with the air intake system. They can also all be caused by several other malfunctions (e.g., the choke is still on), but the quickest, simplest place to start the diagnostic procedure is with the air intake.

Check the pre-cleaner, air filter, and air intake to the carburetor. If the visible parts of the air intake, choke valve, and associated hoses and clamps all appear clean and functional, do a little minor disassembly to check that pre-screener, filter, and line are not dirty or clogged by dirt and debris. If these are clean and open, the carburetor should be getting enough air and the problem lies elsewhere.

Loud noise from the exhaust system indicates a malfunction. Check the components in this order:

1. **Check the muffler by visual inspection.** The bolts or clamp that holds it to the engine may be loose, the muffler may be rusted, or it may have fallen off. New mufflers are usually inexpensive and simple to replace. Replace worn or rusted clamps and bolts as well.

2. **Check the exhaust pipe.** Look for holes due to rust, and for loose connections to the manifold or muffler. Replace if indicated.

3. **Check the manifold gasket.** Look for a sooty spray mark on the engine block around the periphery of the manifold.

4. **Check the manifold.** Look for cracks and/or soot spray.

Problems with the Coolant System

Coolant system problems are indicated by an engine that frequently overheats, or that warms up very slowly. If the coolant level is correct, the radiator screen is clean, and there are no obvious leaks due to failed hoses or clamps, then check the radiator, fan belt, thermostat, and water pump. *Always be careful of a hot radiator.* Opening the cap may release pressure and trigger a scalding boil-over.

Check the radiator for exterior leaks and debris by visual examination. Look for interior plugs by taking the cap off when the engine is cold, and looking down at the radiator core for deposits and/or corrosion. A radiator that is clogged due to hard water deposits or rust will need to be re-cored or replaced, which are generally jobs for a professional.

Check the fan belt for wear, slippage, or breakage. The fan belt is relatively simple to locate, since it's right behind the radiator fan. Replace if indicated. Look at the generator belt too, if there is one, since this may be what turns the fan belt on older tractors.

Check the thermostat. If it's not the radiator or fan belt, the next most likely culprit is the thermostat. A thermostat that is frozen shut causes the engine block to overheat; a thermostat that is frozen open causes the engine to warm up very slowly. Replacing a thermostat is simple, involving taking the old one out and dropping the new one in (with a new gasket); but getting to it to replace it can take some time. This job is well within reach of the amateur mechanic.

Check the water pump. If none of the above are causing the problem, the water pump may have failed. This is bolted on the front of the engine block, where it receives coolant that is exiting the radiator to be circulated around the engine block, then past the thermostat back to the top of the radiator.

Engine Problems That Have Other Causes

If all engine systems check out, the malfunction probably is within the engine block and calls for professional help. Clicking, clattering, slapping, clanking, and knocking noises can variously indicate malfunctions in cylinders, valves, rings, gaskets, lifters, pistons, crankshaft, or associated gaskets, bearings, and gears.

POWER TRANSMISSION PROBLEMS

DRIVETRAIN PROBLEMS ARE CAUSED by the mechanical failure of one or more of the parts that transmit power from engine to working parts: belts, chains, clutch pedal, clutch, flywheel, differential, other gearboxes, axle, driveshaft, and associated parts.

Check belts, pulleys, chains, and sprockets. If, after checking the parts as listed below, there appears to be no malfunction, then the problem is upstream or downstream in the system. Overloading equipment causes frequent problems with belts and chains.

For belts and pulleys:

- Squealing or a burning smell indicate a belt that's stopped due to overloading of the machine or a lockup downstream, so the problem is elsewhere.
- Slippage and lack of power indicate that a belt is loose; this is most often caused by a worn or stretched belt, or by the wrong size/shape of belt, and cured by replacing the belt. Slippage can also be caused by contamination with dirt, oil, or grease; if so, then clean up the pulleys and belt.
- A belt that jumps or pops off the pulleys may be worn, or the pulleys may be worn or misaligned.
- A belt that breaks frequently indicates a wrong-sized belt, or a problem with the pulleys or belt guides.
- Check the manual to see how pulleys and springs should be aligned, and manually turn the pulleys and working parts to see which might be overloaded or locked up due to bearing failure, wear, or some other cause.

For chains and sprockets:

- Chains stretch with use; if one appears to have excessive sag between sprockets, see if it can be tightened using the tensioner guides or sprockets. If not, remove a link; if that is too much tightening, then add a half link.
- You can also hold a chain up sideways to see if it sags a lot; that indicates heavy wear and you may want to replace the chain.
- Chains that pop off sprockets may be due to a loose and/or worn chain, worn teeth on the sprockets, a poorly secured and wobbling sprocket, or a misaligned sprocket.

- A chain that stalls or breaks may be due to wear or to a lockup downstream, perhaps in the bearings supporting its shaft.
- Work the sprockets manually to see which of the above might be the case.

Check the clutch. If the clutch is grabbing, chattering, hard to engage, or slipping, this indicates wear or malfunction. Popping out of gear indicates a problem with the transmission or shifting assembly. But first see if it is something simpler: Work the clutch pedal while watching to see if it pushes down as far as it should and springs back all the way. If not, grease or oil as appropriate. If this doesn't make the clutch work smoothly, you should have an advanced mechanic look at the clutch.

Check differential and any other gearboxes. Look first for oil leaks caused by a missing plug, bolt, failed gasket, or cracked housing. If fixing the leak and replacing the oil does not solve the problem, the gears have become misaligned, worn, or failed, and the box should be replaced.

Check axles. A broken axle is obvious. Replace by carefully disassembling the axle from the attached parts, and installing a new one. Look for collateral damage in the axle housings.

Check driveshaft. If the engine runs and the rest of the drivetrain checks out, but no power is being transmitted (and there may be a lot of noise coming from the driveshaft compartment), there is a problem with the driveshaft. On a tractor, this is a job for an advanced or professional mechanic, as it's buried deep in the frame of the tractor.

OTHER MALFUNCTIONS

PROBLEMS NOT ASSOCIATED WITH the engine or power transmission can be associated with steering, wheels, and brakes; the hydraulic system; or mechanisms and working parts on implements that don't have integral engines.

Steering and Tire Problems

If a machine steers (or pushes or pulls) oddly, always start with the tires or wheels.

Check wheels and tires. Steering that is sluggish and heavy can be caused by low or flat tires. Visually inspect tires, and use a tire air gauge to see if the air pressure is correct. If pressure is low, add air and see if it holds. If it doesn't, there is a puncture, a valve stem leak, or a rim leak if it's a tubeless tire. Remove the wheel and take it to a repair shop.

If a wheel is wobbling, first make sure the lug nuts are tight. If they are, check the rim to see if it is bent. If it's not, the problem may be worn bearings. Replace them if indicated. A wheel that locks up indicates failed bearings. If the bearings are good or have been replaced and the wheel still wobbles, the problem may be the bearings are not tight on the spindle shaft, which can be due to maladjustment or a worn spindle shaft.

If the wheel wanders directionally, it may be a worn tie rod end. Wiggle it or turn the steering wheel back and forth to check for sloppiness between wheel and tie rod.

If the rear tires turn in unison, pushing the machine in a straight line, it may be a locked-up differential. Fixing this is a job for a professional.

Check the power steering system. A steering wheel that is suddenly slow and difficult to turn indicates a problem with the power steering system (if the equipment has one). First check the power steering fluid level. If that is okay, use the manual to locate and identify the hoses and other components of the system, then visually inspect for power steering fluid leaks, hose failure, or belt failure. If you can see the problem and get at it, you can probably fix it, but if you can't, seek help.

Check traction. Ground that is muddy or slick causes the wheels to spin and dig in rather than move forward. Too much load on the machine can have the same effect. Wait until ground conditions improve, and/or reduce the load. Sometimes the machine has enough power but is too light to create the solid ground contact needed for the wheels to grab. Add weights to increase traction (this is commonly done on the front ends of tractors, and on some tillage implements like discs).

Brake Problems

Anything that you ride will have brakes, which may fail or lock up. As always, start with a visual inspection of the parts of the system you can see, making sure that no bolts, springs, clamps, or other small parts are broken or missing. If there is no obvious problem there, consult the owner's manual to find out how to adjust the brakes, including safety brakes. If an adjustment doesn't solve the problem, you will need to disassemble the brakes to get access to the functional parts.

OVERLOADING EQUIPMENT, as shown here, going too fast for conditions, or using a machine for a purpose it wasn't intended for are all common causes of malfunctions and accidents.

Hydraulic System Problems

A hydraulic system that is balky or slow is a common problem. If the fluid level is good, the filter is clean, and the fluid not milky or dirty looking (indicating contamination with water or grit), then do the following.

Check for air in the system. Do this by starting the engine and pumping the hydraulic levers back and forth a few times. If the movement smoothes out, this was the problem, as the bubble has cleared back to the reservoir.

Check for leaks. If the fluid level seems to constantly drift downward, search for slow leaks around the couplings, which are very common. Clean thoroughly, then wrap Teflon tape around the threads of screw-on couplings. If this doesn't work, the couplings should be replaced. Quick couplings that leak constantly should be cleaned; if that doesn't fix them, replace them.

Check the cylinder. If fluid and couplings are good, but the cylinder shaft is balky, the problem is probably in the cylinder. The cylinder can be detached and taken to a shop for repair or replacement.

Check the pump. If only one cylinder or hydraulic function is impaired, the pump is probably okay. If the whole system isn't working, it's probably a pump malfunction.

IMPLEMENT PROBLEMS

AN IMPLEMENT THAT plugs up, locks up, or disengages does so most commonly because it's being run too fast, under too much load, or in inappropriate ground conditions. Other major causes of problems are a lack of grease, maladjustment, and worn parts.

Since many implements are mechanically simple, worn parts are generally (not always) easy to spot and repair. More complicated implements, like hay balers, have more complicated mechanisms. Some, like knotters on small square balers, generally require professional attention, while others, like a slip clutch, can be repaired by the amateur.

Check the slip clutch. A slip clutch that disengages frequently or prematurely even though the load is appropriate for the implement should be checked first to see if it is overgreased. If so, it should be disassembled and cleaned. If this is not the problem, check the owner's manual to see how it should be properly tensioned. If the tension is not the problem, check the clutch cogs, springs, and other parts for excessive wear. Replace the clutch if indicated.

CHAPTER SEVEN

REPAIRS

Repairs happen. With a good maintenance program and close attention to how equipment is running, most of the time repairs will be minor, though they'll still occur — sooner or later everything wears out or breaks.

Disassembly and Reassembly

One of the most frustrating parts of making a more complicated repair is reassembling everything and finding you've done it incorrectly. To avoid this common mistake, take a photo before taking apart anything more complicated than a couple of bolts and a cover. Or, as you disassemble parts, draw a diagram of how things fit together, in what order, and which sides face out. Lay the parts out in order and turned the right way on a clean tarp or worktable, where they won't be disturbed.

GENERAL PROCEDURES

ONCE THE PROBABLE CAUSE of a malfunction has been identified, fixing it is a four-step process:

1. Get to the problem by removing any bolts, covers, and other parts between your tools and the malfunctioning part.

2. Adjust, fix, or replace the bad part.

3. Reassemble everything correctly.

4. Run the equipment to see if it now works.

If this doesn't solve the problem, reexamine the part to make sure it's running correctly. If it is, look for another source of the problem.

REPLACING PARTS

FARM EQUIPMENT IS BUILT TO be durable and functional and has few if any decorative parts. For this reason, it's important not to replace one part with another of a different size, shape, or hardness before first considering how that will impact the way the machine functions. For example, bolts that hold pieces of a frame together can usually be replaced with anything that fits the hole, but other bolts may be of a specific hardness, length, and head shape to accomplish a specific purpose and should be replaced only with the same kind.

When possible, buy equipment of a brand that has a dealer in your area, so that replacement parts are quick and simple to obtain. When ordering parts at the dealer or directly from the manufacturer by phone, have the make and model of the equipment ready; this simplifies the process of identifying the needed part. If you're headed to the dealer's to get a part, take along the owner's manual and the broken part (if it's small). For old tractors and implements, it may be worthwhile to check salvage yards and used-equipment dealers for parts; this can save money.

Repair or Replace?

When looking at a major, expensive repair, get estimates for the cost of the repair and compare these to the cost of replacing the equipment. Factor in also how many more seasons the equipment is likely to last. All equipment eventually gets to the point where age and wear have increased the frequency and cost of repairs to the point where you're better off buying a replacement, and it's too easy to get into a cycle of putting more money into equipment than is justified by its worth or overall functionality.

Old equipment can be scrapped for its metal value or sold to a salvage yard for its parts value.

FIELD REPAIRS

EQUIPMENT DOESN'T BREAK when it's sitting in the shed; it breaks in the field, often when there's not enough time to get equipment back to the shop, run to town for a part, make the repair, and still finish the task before bad weather moves in or a marketing deadline passes.

For this reason, it's handy to have not only commonly needed spare parts on hand but also a mobile tool kit, either in a box bolted to the tractor frame or in the cargo area of a pickup truck or ATV. This allows you to get basic tools to the equipment and make minor repairs quickly. The minimum field tool kit is a hammer, large pair of pliers, flat head and Phillips head screwdrivers, and a couple of sizes of adjustable wrenches.

If you have more room, add a grease gun, penetrating oil, pry bar, shop rags, and whatever else seems to be needed regularly; we routinely carry a set of socket wrenches. Also include some zerks, cotter pins, hitch pins, nails, nuts and bolts, and pieces of wire for jury-rigging emergency repairs; once the equipment is back in the shed you can remove the jury-rig (such as a nail for a cotter pin) and put in the proper part.

BASIC REPAIRS

SOME OF THE MOST COMMON TYPES of repairs are described below. By the time you've done a few of these, you'll be ready to tackle more involved repairs, especially if you have an experienced relative or neighbor to give advice or lend a hand.

Frame and Contact Parts

Worn-out and rusted-through parts should be replaced. If a bent part can't be straightened on the farm with a hammer and anvil or by putting one end between two anchors (such as a bench vise or a pair of close-set tree trunks) and applying force to the other end, call a repair shop to inquire if they think a repair is feasible. If not, replace the part. If the part plays a major role in keeping the equipment properly aligned and functional, and bending it back could weaken it significantly, you are better off replacing the part.

Cracked frame parts should be welded as soon as the crack is detected. If you don't weld yourself, you have the choice of either hauling the damaged equipment into a welding shop or finding someone with a mobile welding unit who will come to the farm.

CONNECTING PARTS

Replacing bolts, gaskets, clamps, hoses, pins, and similar types of connecting parts is self-explanatory. Rivets require a riveting tool, one end of which is used to force out what's left of a broken rivet and the other end to hold the new rivet straight and apply pressure to the end of the rivet shaft to flatten it out and secure it. Read the directions on how to use the tool, and be sure to take time to get everything aligned and secured before installing the new rivet. If the new rivet is loose, remove it and put in another one.

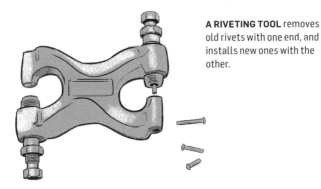

A RIVETING TOOL removes old rivets with one end, and installs new ones with the other.

Springs can be fussy to replace if there are no tensioning bolts or screws to loosen one of the attachment points. In that case, a spring must be stretched far enough to hook over the second attachment point. Try looping a piece of wire through one end of the spring to pull it taut and guide it over and onto the second attachment.

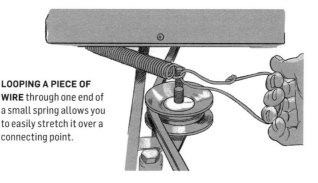

LOOPING A PIECE OF WIRE through one end of a small spring allows you to easily stretch it over a connecting point.

CHANGING A FLAT TIRE

To CHANGE A FLAT TIRE, park on a level spot, and block the other tires with wood blocks. Then loosen each lug nut slightly, while ground friction holds the tire.

Put a jack under the axle, and raise the tire a little higher than necessary off the ground. Pile wooden blocks next to the jack and under the axle, then release the jack so the axle rests on the blocks and the tire is still off the ground. Unscrew the lug nuts, and remove the tire. To replace the repaired tire or put on a new one, position it on the lug bolts, then tighten the nuts by making a few turns on each nut in succession, so they are screwed down evenly and the tire rests flush against the drum or backing plate. Once the lug nuts are snug, remove blocks and jack. When the tire is back on the ground, give a final tightening turn to each lug nut.

Replacing Bearings

When a wheel is wobbling on the axle and the lug nuts and spindle nut are tight and the spindle is not worn, the bearings are probably at fault. Repacking bearings was described in chapter 5; if the bearings are worn, they should be replaced with a duplicate unit. Sealed bearings are always replaced rather than repacked; they are recognized by the rubber gasket between the raceways and because they are bound tightly together. Unsealed bearings come apart.

Rear tractor tires are so large and heavy they can't be repaired or changed without special equipment, so most farmers call a mobile repair service. If you do have a flat rear tractor tire, park the tractor so the leak is at the top; this limits how much calcium chloride solution will leak onto the grass and kill it.

FUEL SYSTEM REPAIRS

IF THE ENGINE SEEMS not to be getting fuel, turn the fuel line shutoff valve to "off," then remove the clamps that connect the line to the gas tank and carburetor. Open the valve, and see if fuel drains at a decent rate into a container. If not, clean out the line, and any screen in front of the carburetor intake, or replace the fuel filter. Reattach the line, and turn on the fuel valve. If the engine is still not getting gas, the problem may be in the carburetor.

When cleaning or adjusting a carburetor for idle speed and richness of the gas/air mix, it's best to use a detailed instruction guide or seek the help of an experienced person. Any carburetor disassembly will require a new gasket kit to avoid leakage. Note that working on a carburetor will often void the warranty on a new engine.

ELECTRICAL SYSTEM REPAIRS

IF A BATTERY IS NOT DELIVERING a charge to the electrical system, clean any debris off the top of the battery, since this can cause a short between the terminals. Then clean any corrosion off the terminals and cable clamps with baking soda and water, or by sandpapering the contact surfaces. When they're clean and bright, apply a light coating of dielectric grease or light lubricating oil, reinstall the battery cables, and see if the engine will start.

If it doesn't, charge or jump the battery, and try the engine again. If the battery won't hold a charge, then replace it. This is a simple and common repair: Loosen the nuts on the bolts that hold the battery cable clamps tight on the terminals, then gently wiggle the clamps off the battery terminals. Take the battery to a farm store or battery store, buy a new one of the same size and power rating, turn in the old one, and install the new one. Be careful of any leaking battery acid — this is hydrochloric acid and will burn holes in clothes and skin.

If the battery is good but the charge is being drained, there's a short elsewhere in the system. Wiring and connections can cause a short because of looseness, corrosion, breaks, or frayed or gnawed insulation letting a naked wire contact metal. New wire and connectors are cheap and not difficult to replace, but it can be a long job finding the short circuit.

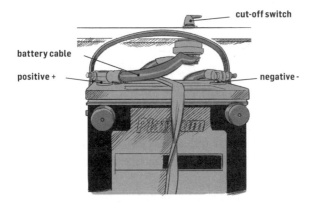

A CUT-OFF SWITCH can be installed between the positive cable from the battery and the rest of the electrical system, to prevent small hidden shorts from draining the battery of charge when the engine is not running.

Other Electrical Components

Replacing a bad spark plug is simple; pull off the spark plug wire, then use a spark plug socket on your socket wrench driver to unscrew the plug and install the new one. Bad spark plug wires and distributor caps are simply removed and replaced by hand. On engines with multiple plugs, be extremely careful to put on the spark plug wires in the exact order they came off both the plugs and the distributor cap. Put masking tape tags on each wire and number them before removal, or just remove one wire at a time. The rotor under the cap easily slips on and off a rotating shaft, also making this simple to replace when faulty. The points are an interrupting switch that sit near the rotor at the base of the rotor shaft, and should be replaced if you're replacing the rotor. The gap space on the points should be set according to the manual's specifications at the most open position.

Replacing failed ignition switches, distributors, coils, generators, and alternators is well within the reach of the farm mechanic, as is adjusting the timing (you'll need a timing light), if you find a good set of directions for that piece of equipment, and perhaps some advice or hands-on help from a more experienced mechanic.

COOLANT SYSTEM

THE MOST COMMON PROBLEMS WITH a coolant system are a failed radiator cap, which is easily replaced; a blown hose; and a failed thermostat. Hoses and thermostats are simple to replace; the hard part is disassembling things to get at them. Be sure to purchase and install a new gasket if you put in a new thermostat.

HYDRAULICS

WE DISCUSSED REPLACING hydraulic hoses and connections in chapter 5. Hydraulic cylinders fail only rarely, and smaller ones are not expensive to replace. If there's a repair shop in your area that deals with cylinders, you may be able to have yours repaired rather than replaced, but first make sure that is the cheaper option. To put on a repaired or new cylinder, unscrew the hoses to remove the old cylinder, attach the new one, then add more fluid to the system to fill the new cylinder. Start the engine, and pump the levers a few times to work air bubbles out of the system, then recheck the fluid level and top off as needed.

Replacing a Bad Muffler

On old tractors, mufflers are usually held on with a clamp or a similar system, so that replacing one is a simple matter of removing the old, buying a new one, and putting it on. It's not necessary to have a specific make of muffler; it just needs to fit on the exhaust pipe.

CHAPTER EIGHT

GETTING PROFESSIONAL HELP

A constant dilemma when you're looking at more than a minor repair is whether to fix it yourself or hire it out. To help determine the most cost- and time-effective answer, call a shop (or two) and ask two questions: first, what is their estimated shop time to fix the problem, and second, does this job require specialized tools.

Budget, weather, deadlines, and complexity of repair can complicate a decision whether to do it yourself or hire it out. This is where experienced neighbors, friends, and relatives can really help with a good perspective on which option will be the best.

WHEN TO GET PROFESSIONAL HELP

IN SOME SITUATIONS, it is more cost- and time-effective to have a professional do the job:

- If something is wrong, you've looked it over, and you're still completely in the dark; for example, an elusive electrical short or an engine malfunction inside the block

- If you have repeatedly made the same repair and the same part keeps breaking down (indicating the problem is elsewhere in the equipment and may be difficult to locate)

- If a repair is going to be physically too demanding or require special tools that you don't own and can't rent; for example, when a tractor body has to be "cracked" in half to get at the drivetrain clutch or a rear tractor wheel has to be repaired or replaced

- If you don't have the technical skills to make a repair; for example, fixing the knotter on a square baler, or welding a cracked frame

- If you're under severe time constraints, such as needing to get a harvest in or a planting done, and just don't have the time to make the repair

WHERE TO GET ASSISTANCE

FOR REPAIRS THAT SEEM DOABLE if you just had a little more time or experience, ask a relative, neighbor, or friend with a good reputation as a mechanic for advice, opinions, or help, and you may be able to do it yourself. For repairs clearly beyond your abilities, take the equipment to a shop with a good reputation that specializes in the type of work you need done.

Jobs that are commonly done by professionals include welding; radiator recoring or replacing; rear tractor tire repair and replacement; drivetrain repairs, including clutch work; differential gear and gearbox repairs; rebuilding generators, alternators, and starter motors; and anything inside the engine block, including failed cylinders, pistons, head, and head gasket.

Finding Good Help

Mechanics run the gamut from excellent to poor, so it's worth your while to inquire among friends and neighbors to see who they recommend. Equipment dealers usually have an on-site repair shop, and most small towns have a selection of stand-alone shops for welding, small engines, radiator work, or some combination of these. Often a local farmer or rural resident will run a repair shop on the side; these sometimes offer better service and more reasonable rates than a branded and dedicated shop.

RESOURCES

Deere & Company. *The Operation, Care, and Repair of Farm Machinery.*
1927–1957.

This handy little hardcover went through 28 editions before publication ceased in the late 1950s. If you can find a copy, it's invaluable for its discussions of how to adjust old-model tractors and implements to run most efficiently and effectively.

Kubik, Rick. *How to Keep Your Tractor Running.* Motorbooks, 2005.

This well-illustrated manual walks you step-by-step through basic maintenance and repair jobs for old tractors.

Naval Education and Training Program Development Center. *Basic Machines and How They Work,* rev. ed. Dover Publications, 1994.

Originally published in 1994 as *Basic Machines* by the US Government Printing Office. This little gem clearly discusses the physics and math behind the mechanisms used in equipment.

Welsch, Roger. *From Tinkering to Torquing: A Beginner's Guide to Tractors and Tools.* MBI Publishing Co., 2005.

One of several books on the topic of restoring old tractors by this author, it offers a lot of excellent information, and a lot of chuckles.

Standard Metric Conversion Formulas

1 liter = 0.2642 gallons = 1.06 quarts

1 gallon = 3.785 liters

1 quart = 32 ounces

Maintenance Log

	DATE	MACHINE	SERVICE	NOTES
Example	9/15/15	big brush mower	replaced drive belt, chg oil, greased, cleaned air filter, checked spark plug, oiled chain, power washed, Sta-bil in tank	order replacement belt

Maintenance Log

DATE	MACHINE	SERVICE	NOTES

Maintenance Log

DATE	MACHINE	SERVICE	NOTES

INDEX

Page numbers in *italic* indicate illustrations.